The Electrification of the
Soviet Union

Also by Craig Raine

THE ONION, MEMORY (OUP)
A MARTIAN SENDS A POSTCARD HOME (OUP)
RICH

CRAIG RAINE

*The
Electrification
of the Soviet
Union*

faber and faber
LONDON · BOSTON

First published in 1986
by Faber and Faber Limited
3 Queen Square London WC1N 3AU

Filmset by Wilmaset Birkenhead Wirral
Printed in Great Britain by
Redwood Burn Ltd Trowbridge Wiltshire
All rights reserved

British Library Cataloguing in Publication Data

Raine, Craig
The electrification of the Soviet Union.
1. Operas—Librettos
I. Title
782.1'2   ML50

ISBN 0–571–14539–6
ISBN 0–571–13958–2 Pbk
ISBN 0–571–14540–X Limited

'. . . and the neat man
To their east who ordered Gorki to be electrified . . .'
W. H. Auden

'Next, he introduced electricity to Ethiopia, first in
the Palaces and then in other buildings . . .'
Ryszard Kapuściński

Preface

The 'Sirens' episode of *Ulysses* supplies the main reason why opera libretti are seldom published without the music which is their justification. Joyce was acutely aware, as an opera lover, of the subordinate role assigned to the words. They are the composer's factotum, to be ordered and reordered as the music requires, much like the boots of the Ormond Hotel: 'to the door of the diningroom came bald Pat, came bothered Pat, came Pat, waiter of Ormond.' Always excepting da Ponte and Hofmannsthal, the librettist may be essential, but he is essential only as a flunkey. Even an opera house needs a lavatory. On the other hand, we tend not to mention it, so perhaps I should volunteer a few words of explanation for asking my readers to consider the plumbing in isolation.

Granted this generally accepted state of affairs, a hierarchy whose base is the librettist, I was only mildly intrigued when Nigel Osborne, the composer, proposed that I should adapt Pasternak's novella for the operatic stage. He had already set some of my poems – 'Professor Klaeber's Nasty Dream' and 'The Fair at St Giles' for the opening of the new Oxford Music Faculty building, and the ten love poems which constitute the first part of *Rich*. My part in this had been the untaxing one of granting permission. Now, he was initiating something more strenuous. However, I agreed to look at *The Last Summer* which he had chosen for its 'symphonic' potential. The novella is complicated by flashback, obscurity, lack of plot and a plenitude of trivial event. Moreover, it exists only in a translation which plumps for literalness where comprehension is wanting. In a word, it looked impossible. My first solution was to travesty the original, retaining only those elements which appealed to me. And I added as much as I subtracted, in the process of producing a treatment with three acts of two scenes each.

At Glyndebourne, I realized, to my dismay, that opera was a communal activity. In our early discussions, we canvassed the names of various possible directors. Peter Brook, it was thought, might well be interested. Clearly, Brook would attract an audience who might be unwilling to make the journey to Glyndebourne merely to watch the outcome of a collaboration between myself and Nigel Osborne. Even so, I was relieved when Brook finally decided not to participate: I had read various accounts of Brook in action and, however brilliant the results, I wasn't looking forward to rewriting and reconceiving my work right up to the dress rehearsal.

In the event, at the initiative of Glyndebourne, Nigel and Peter Sellars met – hit it off – and Peter joined the project. We gathered – reluctantly on my part – for five days at Glyndebourne to discuss the shape of the opera. The first thing suggested by Sellars was that my treatment should be scrapped entirely and that we begin again with Pasternak's text, going through it word by word, even consulting the Russian. I agreed, while pondering how exactly I would convey my decision to resign from the project. In fact, I found Sellars a delight to work with: we argued vigorously, caricaturing each other's positions and sharing a taste for brutal irony. His conception of what we might do was somewhat weakened by the appalling American translation on which he based his ideas. His contribution nevertheless was crucial: he explained just how he would stage the adaptation, using video, so that the libretto would resemble a film script with quick cuts, rather than the equivalent of a modest terrace house architected by Pinero. This agreed, he fought hard for complication, while I was prejudiced in favour of form and simplicity. To my surprise, I found that I enjoyed the group activity and I left Glyndebourne with a sketch of the plot and, more important, a sense of freedom.

Several months later, I began work on the opera, confident that Peter would solve any problems of staging – indeed, welcome them. I knew, too, that our sense of trust was mutual and, without a qualm, I dropped scenes I thought inessential and even a character here and there. Other characters were given less important roles than we had originally envisaged.

At this stage, I still regarded my contribution as subordinate – to the composer, obviously, but also to my own work as a poet. I was, as it were, merely renting out my literary equipment. Given this cast of mind, it was some time before I properly realized what I was doing. For example, I had intended to write the recitative in prose, reserving poetry for the songs. This rapidly became unsatisfying. I felt the need for some formal constraint, since the opera was fluid enough from scene to scene. On the other hand, I wanted the dialogue to be flexible and naturalistic, rather than a minuet of end- stopped lines. My guide here, though not my precise model, was Ben Jonson, whose play *The Alchemist* begins:

> FACE: Believe't, I will.
> SUBTLE: Thy worst. I fart at thee.

In the edition of 1612, this racy exchange can be seen clearly for what it is – a remarkable line of blank verse. Jonson's vivid demotic corresponded to what I was after: a verse line capable of accommodating realistic material as well as more lofty sentiment. The pentameter, however, is a tainted medium. Too many great writers have used it for the drama, so that any predisposition to linguistic richness entails the presence of Shakespeare and the great Elizabethans. Shelley's *The Cenci* or Coleridge's *Osorio* exist as warnings: throughout it is difficult not to hear Shakespeare like a ghostly prompter, speaking just before the characters. (Frost's blank verse line is, of course, the great exception. It is the voice of America – relaxed, idiomatic, understated, garrulous on occasion, tersely eloquent when he chooses.) The bare blank verse line is impossible, too, because Wordsworth has made it his own, as Tennyson's 'Dora' graphically illustrates. I could see why T. S. Eliot had decided on a three-stress line with two of the stresses after the caesura, while allowing himself any number of syllables and freedom to vary the position of the caesura. Eliot's solution, however, was his solution, not mine. Moroever, he had found that the technical challenge tended to oust the drama. He composed poetry, rather than dialogue.

By chance, I hit on the notion of an octosyllabic line, using half-rhyme – and full rhyme when it seemed dramatically required. This line imposes economy on the writer. There is no place for superfluous adjectives and its intrinsic terseness suits modern speech. Moreover, it comes without associations. In it I discovered what I needed. It was both shapely and colloquial.

As for the songs, I had always had the idea that each character would have his own style. In practice, this meant a chance to use metres and rhyme schemes that I don't normally employ in my own poetry. Dramatic necessity required that the songs should not be poetic outbursts insulated from the rest of the drama. I saw my task as the creation of songs that these characters might sing if we lived in a world where people did actually burst into song.

Serezha, who is a poet, was the only character for whom my own way of writing might be appropriate. My poetic voice and his could be analogous. Even here, though, I found myself insisting on the difference at the margin.

Serezha clearly has an earthy side with which I could easily identify, but he is also exalted as adolescents frequently are. There is an element of the 'poetic' about him, as Pasternak clearly realizes, and this can be admirable while, at the same time, susceptible to irony. In the poetry I have written for him, I have tried to do justice to his character as Pasternak presents it. He is obviously gifted because he is an early version of the author, but he is also given to cloudiness and generalized sentiment. His songs attempt to suggest both real achievement and a certain immaturity, particularly in the song for Mr Y. Serezha's version of Mr Y is how, in an ideal light, he sees himself. The song there reflects the element of hubris and over-reach in his self-characterization. His song after Anna has left and his eulogy of Sashka's body are a different matter: the point at issue is not one of self-presentation, but of real self-expression. Mr Y is a testimonial intended for public consumption. The other songs are essentially private and spontaneous. The song about his feelings for Anna is authentically exalted, yet precise, and sung to himself. His song for Sashka is true to his character as we see it in action: it is poetic and earthy, just as Serezha in the preceding dialogue has shown himself to be sensitive but also a man whose appetites have brought him to a prostitute.

Pasternak has only one song in the opera and that was originally written for Lemokh – before we decided that, for dramatic reasons, Lemokh should never speak, let alone sing. Lemokh is the archetypal *apparatchik*, the pragmatic functionary, a man of few words, who characteristically expresses himself through action. As a party member, it is proper that he should have no words of his own. It follows from this that the song, which I had imagined would be set to martial music, is now more ambiguous. The rejection of ordinary life for large political ideas is tempting in one way; in another, it carries in it the reasons for its repudiation.

Hence Pasternak's action of tearing up what he has written. This tearing is justified by the parallel destruction of the Frestln photographs at the end of the opera. The first tearing is an unargued stage-event which is later confirmed by the fate of the Frestlns: by the end of the opera, the *rejection* of family life for politics is shown more accurately as the *destruction* of family life. In any case, we have already seen that Serezha, though a dreamer, is also capable of ironizing his sister's political kitsch – by parody as well as direct refutation.

Natasha's two songs are written for two different sides of her character – the gushing idealist and the opinionated radical. The latter, of course, is closely related to the bossy sister whom Serezha successfully resists. Anna's songs aim at passionate simplicity: she is not a poet, but there is poetry in her predicament and the accuracy of her memories. Like Fardybassov, she is a person familiar with catastrophe, and the power of that catastrophe to move us should, as far as possible, rely on an appropriately simple style. She doesn't wear her art on her sleeve. All the same, within the rhetoric of plainness I have given her, the songs should have their piercing moments.

I have written an opera, but tried to avoid the silliness which sometimes seems endemic, and which Joyce so effectively sends up. In other words, I have attempted to write a serious poetic drama – a work which only differs from other dramatic verse by the necessary inclusion of songs. The songs are not self-contained. They arise out of the drama and out of each character singing them. Accordingly, I offer this libretto to readers as a dramatic work and I would like to record my fundamental disagreement with Auden, another librettist, who wrote:

> on the other hand, the librettist need never bother his head, as the dramatist must, about probability. A credible

14

situation in opera means a situation in which it is credible that someone should sing. A good libretto plot is a melodrama in both the strict and the conventional sense of the word; it offers as many opportunities as possible for the characters to be swept off their feet by placing them in situations which are too tragic or too fantastic for 'words'. No good opera plot can be sensible for people do not sing when they are feeling sensible.

As an account of opera's potential, this is as coarsely thought-out and as lazily generalized as Dr Johnson's famous refutation of Berkeley's idealism. Both men are booting a stone. Nothing which Auden promulgates is a necessary feature of opera. If we applied his general assumption to Shakespeare, we would be left with the statement that no poetic drama is sensible because people do not speak in blank verse when they are feeling sensible. The conventions of opera are as acceptable as those of poetic drama – but the requirement is that they should be used seriously if opera is to rival the drama. Auden found this difficult. Having begun from an almost identical position, I find myself thinking that, in the twentieth century, perhaps opera is the only place where poetic drama can now seem obviously natural.

Craig Raine
May 1985

THE ELECTRIFICATION OF THE SOVIET UNION

—

AN OPERA

Music by Nigel Osborne
Words by Craig Raine
Directed by Peter Sellars

Production Note

This note should be taken as a general guide only, subject to change according to the circumstances of particular productions. The opera is in two parts, with an epilogue that takes place some years after the main action. The libretto is conceived not as drama, but as poetic film. The set is identical throughout, except for the epilogue: it consists of an empty stage whose back wall has two doors on either side of three plain windows on to which film is projected to create, for instance, the illusion of a train seen either from the outside (with passengers looking out) or from the inside (with a moving landscape capable of acceleration to mimic the movement of the train through space). At the front of the stage is a low catasta or plinth. It will be clear that the scenes, as in a film, are intercut: this sometimes involves a freeze of action on one section of stage while other action takes place in a different area; sometimes there is a black-out of one stage area. There is a table at downstage right, where Pasternak sits, writing the action that the audience sees.

The chronology of *The Electrification of the Soviet Union* is as follows: Serezha's arrival at Ousolie takes place in the winter of 1916, immediately prior to the Revolution of February 1917. At Ousolie, Serezha remembers the May and June preceding the outbreak of World War One in July 1914. The epilogue takes place in about 1920 or 1921.

THE ELECTRIFICATION OF THE SOVIET UNION

A libretto based on Boris Pasternak's novella,
The Last Summer, and on his poem 'Spectorsky'.

———

Dramatis Personae

BORIS PASTERNAK

SEREZHA SPECTORSKY

LEMOKH (non-singing)

MR FRESTLN
SASHKA'S HUSBAND }

FARDYBASSOV
TICKET COLLECTOR }
FRESTLN SERVANT

ANNA ARILD

SASHKA

MRS FRESTLN

NATASHA

HARRY FRESTLN (non-singing)

Act One

1. PASTERNAK *at his table writing. On the catasta stands* LEMOKH, *lit so that he appears to be a statue: he is immobile, dressed for winter, a scarf covering his face except for the eyes.* PASTERNAK *is wearing a summer shirt, perhaps with the sleeves rolled up, his braces on show, his shirt open at the neck.*

2. *A noise of trains, whistles, brakes, gradually increasing in volume. Train created by the windows, seen from the outside. Hubbub: a crowd appears, from both doors, with luggage. They gather round the desk and* PASTERNAK *stamps their papers. Clearly, this is a society in turmoil. The crowd streams off through both wings and into the audience, until the stage is empty except for* PASTERNAK *and* LEMOKH.

3. LEMOKH *alters position a fraction.* PASTERNAK *sings:*

PASTERNAK
The unknown soldier comes to life.
The days of kings are gone.
The manifesto is his wife,
the struggle is his son.

The breast is abolished today,
the rifle is polished today.
Compared to a rifle
the heart is a trifle
whose orders we must disobey.

History takes to the street,
a weapon in its hand.
The workers are the true élite,
the peasant owns the land.

The mouth is abolished today,
the razor is polished today.
From Finland to Asia,
the gun is our saviour,
the god to whom we must pray.

The nameless of the human race
will stand together and alone,
without a heart, without a face,
till thrones are overthrown.

Love has been put off today,
phone-calls are cut off today:
to wives and to brothers
and countless sweet others,
the party has nothing to say.

The iron comrade comes to life.
His best friend is a gun.
The sharpened sickle is his wife,
the hammer is his son.

4. LEMOKH *descends from pedestal. Exit* LEMOKH.
PASTERNAK *at his desk stops writing and tears up what he has written. He throws the pieces in the air and they become a real blizzard. The sound of a train on an uphill gradient. A whistle. The blizzard clears. Using the windows once more, a train arrives, gradually slowing down. At first we experience it from the inside looking out: a blur settling down to birch forest and then a country station. As it stops with a jolt, we see the train*

22

from the outside: passengers with luggage can be seen in the
windows stumbling forward. SEREZHA *wipes a circle in one*
window and we see him looking out. Enter NATASHA, *his*
sister; they see each other and wave. SEREZHA *and*
FARDYBASSOV *descend from the train, the former with a*
Gladstone bag, the latter with a kitbag. FARDYBASSOV *looks*
vaguely around, sees no one, opens the kitbag and roots out a
pair of boots to replace the slippers he has been travelling in.
While he is doing this, SEREZHA *and* NATASHA *kiss and greet*
each other. Their conversation is not easy.

NATASHA

You're looking well, Serezha dear,
despite – what? – three days on the train.

SEREZHA

And you? You're well? You look the same.
(*indicates* FARDYBASSOV)
You see? I can't escape this war.
The Russian hero. Home on leave.
He's got a medal in that bag.

(*The train pulls out of the station; the windows accelerate, blur,
then show a bare, snow-covered landscape, a railway halt with
the sign of the stop, Ousolie.* NATASHA *and* SEREZHA *watch it
go, then* SEREZHA *resumes.*)

SEREZHA

The wounded are beyond belief.
All those empty trouser legs . . .
You see the soldiers in the streets,
with one demented jacket sleeve
twitching in the wind, and wait
because you have to watch it seethe,

23

cavort, subside, inflate and narrow.
Why?

NATASHA
Talk about the war tomorrow,
when you've had a meal and rested.
(*awkward pause which they both rush to fill together*)

SEREZHA
So how are Pasha and the kids?
Did you get the toys I posted?
(*together*)

NATASHA
Congratulations on your grades.
You always were a clever boy.
(*laughing*)
Yes, the parcel came last week.
And we're all well.
(*pause*)
I want to say
how glad I am you didn't take . . .

SEREZHA
(*interrupts*)
I had no choice. I tried. I tried
to join but failed the medical.
I'm only fitted for one trade –
the useless intellectual.

NATASHA
It would have been a great mistake.

SEREZHA
What? Dying? For my country's sake?

(They continue to argue in dumb-show, their words obliterated by factory noises and the insistent ringing of a telephone. While they argue, FARDYBASSOV, *now in his boots and with his medal pinned to his lapel, hauls his kitbag to* PASTERNAK'S *desk. He is given a written receipt and turns away to enter the factory, whose noise we've already heard.* SEREZHA *and* NATASHA *first freeze, then disappear during black-out.)*

5. *The factory. Machinery is created by shadows on the windows, as are the listening workers. The telephone continues to ring now and then: it will be answered by* NATASHA *in the next scene.* FARDYBASSOV *sings.*

FARDYBASSOV

The moon was open-mouthed with fear,
on the night the *Novik* went down.
The guns were greased, the decks were clear,
the sea a steady frown.

We knelt there ready for action,
sweating in spite of the cold.
Her plates were shifting a fraction
as the engines throbbed in the hold.

We could see a ship on the skyline
like the beam in a Pharisee's eye.
We could hear the fluttering ensign
like panic in the sky.

The silent salvo of their rounds
died like a line of sparks,
and seconds passed before the sounds
had reached us in the dark.

Their shells had hardly exploded,
five hundred yards off spec,
before our gunners had loaded
and the cases bounced on deck.

Have you seen the Northern Lights
a battle can produce?
Have you heard the fizz of cordite
when it eats along a fuse?

Have you felt the spasm of guns?
Or the burn of an empty shell?
Did you know that eight hundred tons
could pulse like a synagogue bell?

6. *Right on cue, the telephone rings, rather feebly.*
FARDYBASSOV *breaks off, nonplussed. General laughter.*
Black-out. In the dark, gradually fading, FARDYBASSOV *sings
the first verse again. Lights:* NATASHA's *dacha.* SEREZHA
taking things out of his Gladstone bag. We are in his bedroom.
NATASHA *still in her coat, exits stage-left, into the
living-room, to answer the phone. The windows show little but
snow, a bright blanket: then, as we hear* NATASHA's
*conversation, the window on the left generates a blot of ink,
which grows and gradually resolves itself into the menacing,
statuesque figure of* LEMOKH, *who stands like a sentry in the
landscape.*

NATASHA

(*off*)

       Natasha here.

(*pause*)

            Lemokh!
       Oh, is that really you, Lemokh?
       Look, can you come tonight at eight?

(*pause*)
>Hello? Lemokh? Yes, to celebrate.
>My brother's here from Moscow.

(*pause*)
>The poet, yes.

(*pause*)
>Yes, yes, I know.
>Your meeting will be over then.

(*pause*)
>No, I insist, Lemokh. At eight, my friend.

(*re-enter* NATASHA, *without her coat, arranging her hair*)

SEREZHA

(*half undressed*)
>I thought the place was empty. Who is that?

NATASHA

>Lemokh. A student of the proletariat.
>He's coming round to eat tonight.
>You wash. I'll get a few things straight.

SEREZHA

>Perhaps we should be introduced.
>You talk. I'll come through when I've dressed.
>Really, Natasha, I wish you'd said
>someone was here. It's just too bad.

NATASHA

(*laughing*)
>Is this my brilliant brother?

(*takes his head in her hands*)
>What is in here? Brains? Or feathers?
>Is this poetic genius?
>No, look at me. I'm serious.
>Us two, we're really quite alone.

27

SEREZHA

Then who was that?

NATASHA

(*long pause*)

The telephone.
The telephone.

(*pause*)

You know: ring, ring;
hello, who's there? I'm listening.

(*pause*)

I was speaking to Lemokh. OK?

SEREZHA

(*smiling broadly and mugging*)

My brains are not themselves today.
They're off their grub. They need some sleep.
Their posture's poor. Their shoulders slope.
I think they need a set of stays
to keep their bulging guts in place.
Their bowels haven't moved for years.
So help me God, they need a nurse.

NATASHA

(*ruffling his hair*)

It's wonderful to have you here.

(*they embrace fondly*)

It's wonderful to stroke your hair.

SEREZHA

Now tell about Lemokh.

NATASHA

I will.
There isn't very much to tell.

He's what you might call practical.
He isn't a dreamer like you.

SEREZHA

You mean he's a schemer like you.
The party that, the people this:
the empty noise of politics.
I know his type – the purist sort,
gun in hand, and hand on heart.
An activist of peasant stock,
a commissar, that's friend Lemokh.

NATASHA

He stands for everything new,
he isn't a dreamer like you.
He talks like a doctor you trust,
like the judge whose judgments are just:
objective,
protective,
a leader whose word you can trust.

He stands for everything new,
he isn't a dreamer like you.
He struggles as everyone must
for the ones who live in the dust,
distressed ones,
oppressed ones,
the lost ones who live in the dust.

He stands for everything new,
he isn't a dreamer like you.
He's broken away from the past.
The old world has spoken its last,
the fleshpots,
the despots,
the old world is broken at last.

He stands for everything new,
he isn't a dreamer like you.
The time for you dreamers is past,
the new world has woken at last:
all power
is ours;
the people have spoken at last.

SEREZHA

(*clapping ironically, improvises an instant parody*)
I look forward to meeting Lemokh,
but first forty winks in the sack.
This old brain has been on the train,
this old brain is feeling the strain,
this old brain
(*comic pause for inspiration*)
has migraine,
(*speaks the next line*)
this old brain wants to lie down.

NATASHA

(*laughing in spite of herself*)
You pig! I'll fetch the eiderdown.

(*Exit* NATASHA. SEREZHA *lies down in his vest, trousers and
stockinged feet. Lights dim, we see the shadowy figure of*
NATASHA *come and cover him with the eiderdown, then exit.
Lights dim further. We hear a brief reprise of* FARDYBASSOV:
*his first verse, with a repeat to end of the second line ('on the
night the* Novik *went down'). On cue, we hear the amplified
noise of a sinking ship – a guzzling, nightmare plughole sound.
Total darkness. Gradually, faint noises of children's laughter,
vague noises of dishes, kitchen preparations, hissed commands.*)

NATASHA

(*off*)

Children, keep your voices down.

(*after an interval*)

What did I say? I won't tell you again.
Your uncle's tired from the train.

7. *The darkness lightens to sunlight under water which, in turn, resolves itself into the dappled shadows of light on leaves. It is summer – the last summer. A child,* HARRY, *is seated at the piano (not all of it needs to be on stage, just the keyboard end). Enter* SEREZHA, *who picks out a tune.*

SEREZHA

You'll get the hang of it, you'll see.
The fingering is easy, Harry.
The left hand has a broken chord.
That's right. You're coming on. Now *hard* . . .

(*Freeze. Black-out.*)

8. *Black-out lightens a little. It is night. Train effect on the windows, seen from outside.* NATASHA *steps down from the train, after handing her suitcase down to* SEREZHA. *Clouds of steam. A porter wheels a great heap of luggage across the front of the stage, masking their initial remarks. We join their conversation* in medias res.

NATASHA

Sometimes I just can't stand you.
What have I done to offend you?

SEREZHA

Nothing.

NATASHA
You should be pleased I'm here.

SEREZHA
I am.

(*pause*)

You're like a questionnaire,
that's all.

NATASHA
I only want to know
you're happy and provided for.
And if you have a job. I've come
two thousand miles to know. Well?

(*pause*)

Damn!

SEREZHA
You mustn't fret about me. I'm fine, fine.
I haven't asked you for a loan,
have I? And, look, you came for fun.
Natasha, sweet, you didn't come
two thousand miles with me in mind:
you came to shop and see your friends.

NATASHA
Where do you live? I want to know.

SEREZHA
Where do I live?

(*pause*)

Moscow. Moscow.

(*Black-out.*)

9. A crush-bar at the theatre. NATASHA *gossiping with friends about* SEREZHA *during the interval of a performance of* Three Sisters. *The windows show photographs of three actresses, who move almost imperceptibly.*

NATASHA

Moscow, Moscow, Moscow, Moscow.
Three sisters with a single whinge.
I only wish they'd up and *go* –
strip off their qualms and take the plunge.

But Chekhov revels in regret,
the sad, the tragi-comic fate:
three sisters whom we soon forget.
That's why his plays are second-rate.

His characters are fatalists,
no politics in any sense.
What we need are activists
who only speak the future tense.

(the photographs in the windows change to Trotsky, Lenin and Stalin – a pause, then all in drag)

But Chekhov's catching like the flu.
My brother's caught a nasty dose
of lifelessness at twenty-two.
It's such an irritating pose.

He has a job at Frestln's place,
the man who runs that magazine:
a member of the chosen race
whose rumoured wealth is just obscene.

33

Frestln. You know, the editor,
Kerensky's friend and advocate.
His wife's a social predator,
but he's not bad. Just out of date.

My brother's tutor to their child.
For once he's fallen on his feet.
The governess, whose name is Arild,
must be the reason he's discreet.

(*Black-out.*)

10. ANNA ARILD, *in mourning, crosses the stage diagonally and exits.*

11. MR FRESTLN *bustles across the opposite diagonal, clutching proofs in a great sheaf. He pauses at* PASTERNAK's *desk, makes a quick correction. Exits.*

12. *Enter* SEREZHA *with* MRS ARILD. *They are already deep in conversation.*

SEREZHA
I hate to see you so distressed.

ANNA
She treats me like a chambermaid.
Her skivvy. Not a governess.
Her, *her*: a Jew, when all is said
and done.

SEREZHA
Now, that's a prejudice.
There's not a drop of Jewish blood
in either one, however rich

34

they are. I know she can be rude,
but he is very generous.

ANNA

And so is she – with her tirades:
'A button's missing from this dress!
My stocking's torn! The bed's not made!
My dressing table's in a mess!
Why have I been disobeyed?'
Because she knows I'm penniless,
I'm like a soldier on parade.

SEREZHA

But don't you see?

ANNA

 See what?

SEREZHA

   She's jealous.

ANNA

Of *what* precisely?

SEREZHA

   Youth. She feels betrayed
because you're young. I'm serious.
Don't laugh. You make her feel her age.
That's why she's so imperious.

ANNA

How sensitive, how very sage,
how very deep,

(*pause*)

   how credulous.

But you are in a special role
as Mr Frestln's protégé:
the house-trained intellectual,
the poet of a page a day,
the favourite.

SEREZHA
You hate me, too,
I see. I only want to help.

ANNA
There's nothing, nothing, you can do.
I don't hate you. I hate myself.

I died the day my husband died.
There's nothing you can do.
I cried the way my husband cried,
the way the dying do.

He turned his head and vomited.
I wiped away the spew.
And when, instead, he soiled the bed,
I cleaned away that too.

His chest was bare. I knew the hair,
I knew the way it grew.
I knew his lips, his fingertips,
but now his lips were blue.

I washed his sheets and walked the streets,
the way the living do.
I hung them out, then walked about,
but I was lifeless too.

I sat there and forgot to blink,
the way that dead eyes do.
I poured his medicines down the sink,
till I was empty too.

And I was brave beside a grave
dug slightly out of true.
The coffin bands slid through those hands
and I was buried too.

I died the day my husband died.
There's nothing left for you.
I played like sunlight on the spade,
and then I faded too.

13. *During the initial conversation between* SEREZHA *and* MRS
ARILD, *servants have laid the table for supper. As* ANNA *ends
her song, the lights, which have been strong sunlight, fade
dramatically to dusk on the words* 'and then I faded too'.
With the change of lights, they part: MRS ARILD *sits at the
piano and* SEREZHA *joins the* FRESTLNS *at the dinner table.*

FRESTLN
And how's the story coming on?
Or have you got a writer's block?

SEREZHA
Right now, I still don't have a plan.
The difficulty is . . .

FRESTLN

(*interrupts*)

        you're stuck!

(*laughs*)

You mustn't force your inspiration.
Just take your time. That's my advice
to all the younger generation:
slow down. You need to find your voice.

MRS FRESTLN
Be careful of the plates. They're hot.
The soup smells good. I always say
that poets should consult the heart,
the source of tragedy and joy.
The works of Pushkin throb with blood;
the eloquence of Lermontov
entreats the soul as poets should.
Confess your heart and speak for love.
It takes the soul and nothing less
to dip your pen in storm and stress.

14. MRS ARILD *at the piano begins to play Chopin violently,
drowning the conversation at the dinner table. A great crashing
of chords, as she brings her fists down indiscriminately, is
paralleled at the dinner table – where the crockery leaps in the
air.* HARRY *screams. General consternation: everyone stands;*
HARRY *is led out by* MRS FRESTLN *with her hand over his
mouth. Black-out as* FRESTLN *blows out the candles.* MRS
ARILD *still lit by a single candlestick. She modulates her rage
into a perky mazurka or polka and* SEREZHA *appears at her
side. He closes the lid of the piano, bows and in dumb show
offers to dance with her. She takes his hand, but as they rise,
they are both suddenly embarrassed by the intimate physical
contact. They stand slightly apart and freeze. Black-out.*

15. *Gradual fade-up to bright sunlight. The windows are a train,
showing a slow procession of birch trees, countryside, a lake. The
train halts:* SEREZHA *and* MRS ARILD *descend. Swaying
leaves and water in the windows.*

SEREZHA

Believe me, I'm your friend:
yes, you can always speak about
the past's eternal ampersands
to me. Those *ands* as long as thought:
'And then he said, and then I cried,
and then he took my hand, and then . . .'
The things we needn't memorize
which live with us until the end.

ANNA

Just held together by that *and*.
O Serezha, you're my friend:
you listen and you understand.

And then we loved, and then he died,
and then there was no then.
My lips would ask, then his replied,
and then, there was no then.

And then he knew his life was past,
and then there was no when.
When every breath seems like the last,
there is no time for then.

With curtains drawn against the dawn,
we waited for the when,
when he would leave me on my own,
in daylight once again.

And then he said he hated spring
with all its sense of when,
when all the buds are promising
that life begins again.

And then at last he hated me
for lovers I would take,
the one who had created me,
whose heart was his to break.

My heart was his alone to break
and when there is no when
all time will stop for our sake,
and then we'll kiss again.

(*As* MRS ARILD *sings about her husband's jealous fears,*
SEREZHA *takes her hand: at the line, 'for lovers I would
take'. For the remainder of the song, she tries to struggle free,
but with the last line she succumbs. A long kiss. Then she slaps
his face very hard and exits. As she leaves, a gradual change of
light takes place – from bright sunshine to a misty dusk to
darkness and moonlight, in parallel with the words of*
SEREZHA's *song.*)

SEREZHA
She leaves and the light
goes slowly blind,
the lovely colours of the lake
begin to fail,
the perfect circle of the sun
contracts into the blackbird's eye.

Soon each sad,
each separate star will shine
like someone on the verge of tears,
hoarding light like hurt.

She leaves but the sun
is mine to keep.
I feel it burning in my chest,
as if a lens
had focused there
this painful pill of fire.

Soon each sad,
each suffering face will fade
and daffodils grow molten in the dawn,
hoarding heat like hope.

I cannot close my eyes.
The sleepless mirror of the moon
is holding her reflection:
a widow watching through the dark.

(*Exit* SEREZHA.)

16. *Moonlight still. Enter* MR *and* MRS FRESTLN *in their
night clothes, on their way to bed.*

MRS FRESTLN

He's fallen for the governess,
you know, that boy. I told you so.
He's all cow's eyes and soppiness.
But what I'd really like to know
is where he disappears to at night.

FRESTLN

(*yawns*)

My dear, as usual, you're right.

End of Act One

Act Two

1. *Darkness. A sound of knocking and the distant noise of shunting trains. Enter* SASHKA, *a prostitute. She is naked, except for stockings, one of which is crumpled round her ankle. She carries an oil lamp and sets it down beside a mirror – a mirror which, in the previous darkness, has held the reflection of the full moon.*

SASHKA
Just wait a moment till I'm dressed.
(*pulls up a stocking and secures it with a garter*)
And cut the noise. I heard, I heard.
(*the knocking stops. She consults the mirror and briskly pitchforks her hair with her hands*)
A girl should always look her best,
but, Sashka, pet, it's sometimes hard,
what with the price of rouge these days.
But still, your face *deserves* a spot.
(*dabs her lips; the knocking recommences*)
And then the cost of knickers, stays . . .
(*breaks off, looks towards the door*)
You'd think this was a knocking shop.
(*laughs, then shouts*)
Knock, knock! You'd think you were the police.
I'm *coming*. Yes. All right, all right.
I only do my best to please.

(*She opens the door. Enter* SEREZHA, *who says nothing. She takes his face in her hands.*)

SASHKA

Ah-hah! Ah-hah! My favourite.
Little Sashka's starving poet.
In every sense.
(SEREZHA *attempts to kiss her and fails*)
In every sense.
Hey! first things first.
(SEREZHA *again tries to kiss her*)
That means your coat.
(*she takes his coat and hangs it up*)
You know I can't begin at once.
It's not polite. A girl must keep
her standards up, if nothing else:
otherwise she just can't cope.
The band tunes up *before* the waltz
or everything just goes to pot.
Which reminds me that . . .
(*she takes a chamber pot from under the bed*)
I need a pee.
(*she squats; meanwhile* SEREZHA *takes off some of his clothes and puts money to one side*)
You're very talkative. What's wrong?
You make a girl feel like a slut,
you know.
(*goes to him*)
That thing in there's a tongue:
use it.
(*he again tries to kiss her*)
I didn't mean like that.
Let's start from scratch. Let's try Hello.

SEREZHA

Hello.

44

SASHKA

That's brilliant.

(*pause*)

The cat . . .

SEREZHA

(*long pause during which he looks puzzled*)

Sat on the rug?

(*pause*)

Could eat no fat? No?
I'm sure that's right.

SASHKA

(*laughing*)

Let's go to bed.

(*They kiss at length.*)

SEREZHA

(*tenderly*)

Hello.

SASHKA

Hello.

(*pause*)

I love your head.

(*They lie down on the bed together.*)

SEREZHA

Your body beside me
ticks like a clock
with your brilliant blood.
I love you in your time machine.

Your broken veins
are galaxies, glowing
with their far-off fires.
I love you in your time machine.

Mackerel shine
in your hips and thighs
like bark on the silver birch.
I love you in your time machine.

Your slow nipples
gather closely in the cold,
dark as rose-hips in December.
I love you in your time machine.

The labia's uncrumpling leaf,
the twelve-tone concerto
for digestive tract.
I love you in your time machine.

The soft cervix
like a peeled banana,
the sphincter's strong anemone.
I love you in your time machine.

These stepping stones
of vertebrae, the navel's
little lotus knot, tidal ribs.
I love you in your time machine.

The smell of you,
of hay-barns, harbours,
the taste of you on my tongue.
I love you in your time machine.

The glaze of youth
is forty-two years old
and cracked with time.
I love you in your time machine.

SASHKA

You're such a funny boy, you know.

SEREZHA

Am I?

SASHKA

Yes. Yes, you are. It's nice.

(*pause*)

How did you know I'm forty-two?

SEREZHA

I guessed.

SASHKA

It shows up in my face.
The rest's OK. My breasts aren't bad.

SEREZHA

I think your face is beautiful.
Your breasts are, too.

SASHKA

They're big. That's good,
isn't it? Most men like a handful.

SEREZHA

I've never asked another man.

(*pause*)

Is it different, or the same,
the way you feel when you're a woman?
I'd like to know.

47

SASHKA
We don't complain,
we girls. You stick out where we are flat.

SEREZHA
That isn't what I meant at all.
I'm serious.

SASHKA
You're gentle, gentle.
Someone else once asked me that.

I was wearing a blue woollen dress
and the boy was a boy like you,
so I gave him my name and address
and I let him feel me, too.

He was young and dying to know
what a girl was like in the flesh,
and I hadn't the heart to say no
when he lifted up my dress.

I remember his careful hands
and the way that my legs were wide,
how I throbbed like a swollen gland
when I felt him feel inside.

I can see the scar on his cheek
and the look that meant he was shy,
or the nose like a kestrel's beak
and his navel's Tartar eye.

But you never forget the first
though you travel to distant parts.
From Odessa to Novosibirsk,
he's the house you know by heart:

you remember with total recall
every hair that was trapped in the paint,
and that tap on the outside wall
or a hinge's one complaint.

But a boy and a blue woollen dress
are enough for a girl to conceive.
When he threw away my address,
home was a place to leave.

So I live in the suburbs now,
where the clients can visit my place
and grunt like a boar on a sow
or come all over my face.

SEREZHA

(*after a longish pause*)
    You think all men are brutes. We are.

SASHKA

Not all of them. Just some of them.
I'm very fond of you,

(*pause*)

                  so *there*.
Let's get your trousers off and then
you'll see how much I fancy you.

(SEREZHA *kneels and* SASHKA *busies herself with his waistband and belt. The door bursts open and, after an interval,* SASHKA's HUSBAND *enters, exits, and re-enters again. He is completely drunk and staggering like someone in a storm at sea.* SEREZHA *makes as if to go, but* SASHKA *puts a hand over his mouth and indicates they should watch the performance. Throughout his 'speech', the* HUSBAND *undresses clumsily, while* SEREZHA *gets into clothes behind him.*)

HUSBAND
Mayday. Rough seas. Abandon ship.
This is your capstan calling all crew:
all hands on dick; all hands on dick.
(*silent convulsed laughter at his own joke*)
Women with chilblains first.
(*delighted laughter ending in a coughing fit*)
                            Oh dear . . .
Man the wifebelts.
(*struggles into a corset of* SASHKA'*s and falls*)
                    Man overboard.
He's in the drink.
(*more wheezing laughter*)
                    He's in . . .
(*convulsed again*)
                            I'll steer.
(*gets to his feet and mimics a helmsman*)
                    Whoops! You'd need to be a whore
to handle all these prongs.
(*loud belch*)

                                Ale-force ten.
Oooh! fire amidships.
(*ruefully pats stomach*)

SASHKA
(*clears her throat and coughs loudly*)

HUSBAND
Pardon.
(*He freezes, then slowly unfreezes, picks up his shoes, and exits
unsteadily, keeping his back to the bed, without another word.*)

SEREZHA
I have to go. I can't stay now.

HUSBAND

(*off*)

Aye-aye, scupper.

SASHKA

No, please stay.

Husbands! I could . . .

(*makes strangling gesture*)

HUSBAND

(*off*)

And there she blows!

SEREZHA

Another night.

SASHKA

I'll see you on the way.

Hang on. I'll just slip on a dress.

SEREZHA

No, don't. You'll

(*pause*)

catch your death of cold.

(*attempts to kiss her goodbye*)

SASHKA

(*seriously*)

Then go.

(*turns away from his kiss*)

(*Exit* SEREZHA.)

SASHKA

(*plaintively to the open door and the room in general*)

But don't you lose my address.

(*long pause, during which she tidies up*)
      I'm forty-two. I'm getting old.

(*Black-out.*)

2. *Knocking, knocking. Enter* SASHKA, *wearing only a peignoir and a pair of felt boots. She is carrying a lamp, as before.*

SASHKA
It's half past two, so just pack up.
Shut it. Or else I'll call the police.
(*She opens the door and uniformed police enter and arrest her.*)

1ST POLICEMAN
You're on the game.

2ND POLICEMAN
(*laughing*)
      We'll knock you up.
(*Black-out. In the darkness, sounds of violence.*)

SASHKA
Please, please, oh don't, *please* . . .
(*screams*)

3. *Black-out still. Silence, then another scream – but this time one of irritation, not terror. It is* MRS ARILD *on whom the lights go up. The Frestln house.* MRS ARILD *quarrelling with* MRS FRESTLN.

ANNA
Ooooh!
(*screams again*)
      I swear to God . . .
(ANNA *moves diagonally across the stage, pursued by* MRS FRESTLN.)

MRS FRESTLN
But Harry's ill:
you can't leave now.

ANNA
Oh, can't I just?

(*mimics*)

But Harry's ill. I hate you all.
I'm sick myself. I need a rest.
My bags are packed, and that's that.

MRS FRESTLN

(*wheedling*)

Look,
Anna, my dear, I understand.
A little touch of overwork . . .
Try to see me as a friend:
you and I must have a talk
and get things straight.

ANNA

(*shakes her head*)

Today,
this afternoon, I'm off.

MRS FRESTLN
You're ill,
you know, you need a holiday.
Come to the country with us. I'll . . .

ANNA
You! You'll exploit me like a slave.

MRS FRESTLN
You're being inconsiderate
and quite unfair.

ANNA

Sweet God above.

(*makes as if to exit*)

MRS FRESTLN

But who will see to Harry? Wait.

ANNA

Serezha Spectorsky.
Why not? That's what you pay him for.

(*Exit* ANNA.)

MRS FRESTLN

If he was here. But where is he?
Where is the precious tutor?

(*Exit* MRS FRESTLN.)

4. *The bright sunlight begins to fade into dusk, then night.*
Sounds of bustle, orders, preparation. Lights then come up on
SEREZHA *in the Frestln drawing room. He pulls a sash-cord to*
summon a servant.

SEREZHA

Ah, Petya. Good. At least you're here.
It's like the *Marie-Celeste*.

SERVANT

Eh? Yer what?

SEREZHA

I mean I was wondering where . . .

SERVANT

They've gone?

(*pause*)

They've gone, that little lot.

SEREZHA
Where to? A visit? Town? Thin air?

SERVANT
The dacha at . . .
(*trails off*)

          Or that other one . . .
(*begins to exit*)
Yes.
(*pause*)

      Master Harry's that unwell,
you see.
(*pause*)

        Last night. And what's-her-name,
the snooty one . . . You *know* . . . the swell,
she's gone too. Left. Yes, quite a storm.

SEREZHA
But where's she gone?

SERVANT
(*pause*)
        Ill.
(*pause*)
        In her room.
(*Exit* SERVANT. *Lightning.* SEREZHA *dashes out. Long roll of thunder. Black-out. Sounds of rain pelting down.*)

5. *Black-out. Sounds of heavy rain continue. A gentle insistent knocking. A lightening to dusk.* ANNA ARILD *lying on her bed. Her room should clearly parallel* SASHKA'S *in its bareness. She is unconscious and doesn't hear* SEREZHA'S *tapping. Finally,* SEREZHA *enters anyway. He goes to the bed, feels her brow.*

SEREZHA

Please. Please. Don't die. I love you. Please.
(SEREZHA *gets out a bottle of smelling salts and waves them
under her nose.* MRS ARILD *begins to revive.*)

SEREZHA

Thank God. I want to marry you.
Say you'll consider it at least.

ANNA

Who's that?

SEREZHA

It's me.

ANNA

Who? Oh, it's you.
Hello.
(*pause*)
I've got a splitting head.
And what's that smell? Ugh, smelling salts.
(*waves them away*)

SEREZHA

Anna. I thought you must be dead,
just lying there.
(*pause*)
I've got my faults,
I know, but will you marry me?

ANNA

It's raining.
(*pause*)
What's the time?

SEREZHA

What?

(*takes watch to the light*)

Eight.

(*takes her hands*)
Anna, I love you. Don't you see?

ANNA

(*still dazed*)
What? See what? Turn up the light.
(SEREZHA *doesn't move.*)

ANNA

(*impatiently*)
What do you want?

SEREZHA
To marry you.

ANNA

(*clear now*)
You're very young. I need to think.
I still feel strange and
(*pause*)

out of true.
(*pause*)
Give me an hour to decide.

SEREZHA

(*fervently*)

Thanks.
(*He tries to embrace her, but she avoids him – getting up so the
bed is between them. For a moment, they look at each other. Exit*
SEREZHA. *Black-out.*)

6. *The Frestln drawing room. In a circle of lamplight, which is a version of the spotlight to come,* SEREZHA *is writing: he is so absorbed that he doesn't notice when* MRS ARILD *quietly enters, with a suitcase and umbrella. She watches him, then leaves.*

SEREZHA

(*writing and mumbling*)
> Uh-huh, uh-huh. It's algebra
> and Oscar Wilde. Er

(*pause*)

> shy, but proud.
> Just one central character,
> the genius, Mr Y.

(*makes note*)

> Great crowd
> of people when he's up for sale,
> with lots of cash.

(*pause*)

> But poetic . . .

(*pause*)

> the bidding like a rising scale.
> Why not?

(*writes*)

> Somehow prophetic.
> He sells himself to save the world.
> Redeems

(*writes as he speaks*)

> the widows . . . prostitutes . . .
> and women everywhere . . . poor girls . . .
> A sale to clothe the destitute.

(*Black-out.*)

7. *Black-out maintained. In the darkness, sounds of activity and excitement. Then repeated cries of 'Mr Y'. Spotlight on*

SEREZHA *in tails like a concert pianist. He sits at the piano and plays – brilliantly. Huge applause. Black-out.*

8. *Lights.* MR *and* MRS FRESTLN, HARRY, SERVANTS, *rushing about with suitcases and trunks, armfuls of clothes. General hubbub. Black-out.*

9. *Hubbub continues. Sudden spotlight on* MR Y. *He is standing on the catasta, downstage centre, and* PASTERNAK *is beside him, holding an auctioneer's gavel.*

AUCTIONEER

(raps with gavel on the catasta)
       Thank you.
(silence falls)
              Presenting . . . Mr Y.
       Philanthropist and genius,
       who will sell himself today.

VOX POPS

Himself? He can't be serious.
Why? It's utterly ridiculous.

AUCTIONEER

Mr Y is doing this
to help all women in distress.
But first
(hubbub dies down)
             but first, a short display.
       Thank you. I give you . . . Mr Y!
*(*SEREZHA *bows and begins Mr Y's song.)*

SEREZHA

I bring you the now
and I bring you the future.
I am God inventing the world.

I change my leafless mind
and stencil catkins
on the frozen hedgerow.

In Africa, to feed,
the elephant furls up
its living ammonite of flesh.

I have invented snows:
soft demerara sugar, pastry,
sorbet, wet salt, cocaine.

The sting-ray stirs,
my effervescent masterpiece,
murky on the ocean bed.

I thought of man,
whose dead live on in dreams,
whose mind flies through the night

with miracles and minotaurs
although his lexicon is drunk –
yet wakes to safety.

I invented the baby
lifting its head
like a lizard in sunshine.

Sublime, I stirred
the centrifugal universe
and watched the stars disperse.

Here, I placed
the thousand shot-silk moons
by the Champagne Gates

and there, the whirlpool wind,
where planets drown
at Spartan Spectros.

I plumb the depths of space
where the broken stars of Vesta
hatch round far-off Hercules.

I bring you the tides of time,
the chariots of hydrogen.
I am God inventing the world.

(*As* SEREZHA/MR Y *sings, the auctioneer draws attention to
the windows which illustrate each stanza. Whereas before the
film has been black and white, it is now in brilliant colour.
Applause. The bidding begins immediately, in a scale. A comic
close harmony – barber-shop quartet.*)

VOX POPS

One million roubles.
Three million roubles.
Five million roubles.
Six million roubles.
Eight million roubles.
Nine million roubles.
Ten million roubles.
Twelve million roubles.

(*Black-out. Silence.*)

10. PASTERNAK *writing at his desk. Slow fade.*

11. *The sounds of riot, civil commotion, sporadic gunfire. Enter*
MRS ARILD *with two hefty suitcases. She crosses the stage and
exits. Enter* SEREZHA *from a different direction, calling her
name, 'Anna, Anna, Anna Arild'. Exit* SEREZHA. MRS

ARILD's *face appears at a window as* SEREZHA *re-enters. He sees her, but the window immediately turns into a train window and rushes her into oblivion. Black-out.*

12. THE FRESTLNS *boarding a train, helped by* SEREZHA, *who remains on the platform. There is a* TICKET COLLECTOR.

SEREZHA

Goodbye for now. Take care. Take care.
(*to* COLLECTOR)
What? I see.
(*to* FRESTLNS)
He says he has to see . . .
(*loud whistle; the* FRESTLNS *pass down the tickets, which are punched and handed back.*
The train pulls out. SEREZHA *waves for a moment longer, then leaves the platform slowly.*)

TICKET COLLECTOR

There's confetti round my shoes,
but I'm no wedding guest.
I'm the one who couldn't choose
which girl I loved the best.

No, I never could decide
which one I loved the most.
I'm the groom without a bride,
a glass without a toast.

(*Black-out.*)

13. *Gradual lightening.* SEREZHA *at Ousolie in his vest and braces.* NATASHA *is introducing him to* LEMOKH, *who is dressed, as before, for winter.*

NATASHA

(*to* LEMOKH)

  It's good of you to come, Lemokh.

  Lemokh. Serezha Spectorsky.

  Serezha. Lemokh.

(LEMOKH *extends his hand, expressionlessly.* SEREZHA *does not take it. Freeze.*)

Curtain

Epilogue

About three years later. The Revolution has taken place and the Bolsheviks are in power – at the stage when local soviets more or less controlled their own territories.

The lights go up on a full set, for the first time. It is the Frestln drawing-room fully furnished with piano, table, chairs, carpets, sofa, knick-knacks, lace doilies, antimacassars, etc. There are curtains on the windows, which reach to the floor. High bourgeois Biedermeier.

Enter FARDYBASSOV *in uniform with a step-ladder. He takes down the curtains, lays them on the sofa, and exits with the ladder.*

For a time, the set is empty. Then we hear a persistent knocking. Enter SEREZHA *rather gingerly. He is older and shabbier.*

SEREZHA

(interrogatively)
          Hello? Hello? Mister Frestln?
(long pause, while he looks around)
          It's like the *Marie-Celeste*. Hello.
(He goes to the window and looks out into the garden. Silently, MRS ARILD *enters. She is in the uniform of a party official, with a gun in a holster. A new severe hairstyle, well-cropped.)*

ANNA

(quietly)
          And who the hell are you? I said
(SEREZHA turns)
          who the hell . . . *You!*

SEREZHA
Anna.

ANNA
(*by way of greeting*)

Comrade.

SEREZHA
I came to see if they . . .

ANNA
They've gone.

SEREZHA
I see. And you?

ANNA
Your jacket's torn.
You're in poor shape. Look at your shoes.

SEREZHA
Yes, I thought . . .
(*gestures towards the room*)

ANNA
You thought the Jews
would help you out.
(*pause*)

They've gone.

SEREZHA
Yes. Where?

ANNA

You'd better just get out of here.
Now, while you can. I'll . . . For Christ's sake,
can't you grasp . . . ? You're so half-baked.
I'll spell it out. We confiscate
the goods of traitors to the state.
The Frestlns were class enemies.

SEREZHA

(*nods*)

Were?

ANNA

Are. Were. What's the difference?

SEREZHA

A difference of sympathies,
perhaps. The Frestlns were my friends.

ANNA

I know. What's more, the Party knows.
The evidence is here on show –
incriminating photographs.
(*She goes to a bookcase and removes a photograph album.
Together they flick through the pages. What they see is projected
on the windows – happy family shots, in which* SEREZHA
features a lot. MRS ARILD *is not in evidence.*)

ANNA

Not one of me. Well, that's a laugh.
But no surprise. She couldn't stand
(*mimics*)
'that frightful governess', could she?
(*She closes the album. Static on the windows are photographs of*
MR FRESTLN, MRS FRESTLN, HARRY. *They remain.*)

SEREZHA

(*smiling*)

No, she could never understand
your

(*pause*)

your special kind of beauty.
Your . . . All that snobbish culture made
her

(*pause*)

blind.

ANNA

You flatter me, comrade.

(*pause, then weakening*)

Serezha, it's time to go.

(*He reaches out to her, but she doesn't take his hand.*)

ANNA

Just go.

(*pause*)

You were a gentle boy.

SEREZHA

Go where? North, south, the album stays.

ANNA

Oh, that. That's easily destroyed.
If I choose.

(*pause*)

For you, I choose.

(*They look at each other for thirty seconds. Then* SEREZHA
leaves. MRS ARILD *turns with the album as* LEMOKH *enters,
with* FARDYBASSOV *and* PASTERNAK. PASTERNAK *sits at
the table, making an inventory.* LEMOKH *is idle while*
FARDYBASSOV *stacks up the furniture. The dialogue here*

depends on what is on stage. FARDYBASSOV *calls out 'sofa' or 'dining chair' as he works and* PASTERNAK *repeats each item by name: 'one sofa', 'one dining chair'. By the time they have finished, the room is in total disorder: a revolution has taken place.* ANNA *has no opportunity to hide the album.*

When everything seems complete, LEMOKH *takes the album from* MRS ARILD. *She has been holding it nonchalantly. He opens it and turns the pages. As he looks, the pictures in the windows tear from side to side, then top to bottom. Then the torn images blur as the windows become a train.*

Freeze.

Black-out.)

Curtain.